# Color Theory for Printing

# Color Theory for Printing

Foundations for Predictable Color in Print

Bob Binder

ISBN: 979-8-9947729-1-1

Published by RGBinder, LLC
Printed in the United States of America

# Table of Contents

# Purpose and Scope

This book explains what color is, how we see it, how we measure it, and how we control it in printing. The goal is not to make the reader a color scientist. The goal is to provide the core principles that make color predictable in real production: on press, in prepress, in proofing, and in quality control.

Color management discussions often jump straight to tools, targets, profiles, and tolerances. Those are important, but they only work well when the fundamentals are understood. Many day-to-day disagreements—proof versus press, shift versus shift, plant versus plant, instrument versus visual—trace back to the same underlying causes: illumination changes, viewing conditions are inconsistent, materials behave differently than expected, or measurement assumptions do not match how standards were created. This book focuses on those root causes so the practical decisions that follow are easier to make and easier to defend.

The approach is deliberately practical. Terms are introduced as they become useful, then defined more precisely as the story develops. Examples are drawn from a range of printing scenarios because the same principles apply whether the color is produced with process inks, spot colors, or expanded-gamut systems, and whether the substrate is paper, film, foil, or another material. The intention is a shared language that helps different roles—operators, prepress, quality, suppliers, and approvers—make consistent decisions using the same definitions and assumptions.

This is a fundamentals book. It does not attempt to be a complete reference on vision science, colorimetry, or ICC profile building. When deeper detail is helpful, it is summarized and tied directly to a workflow decision. The measure of success is simple: after reading, you should be able to predict the most common causes of color shifts, choose appropriate viewing and measurement conditions, and communicate color requirements clearly enough that results can be repeated.

# How to Use This Book

This book is designed to serve as both a narrative guide to color theory and a technical reference for printing professionals. To balance high-level concepts with deep technical detail, I have included two distinct types of content:

The Main Narrative: This follows the core flow of each chapter, providing the essential foundation for understanding color in a printing environment.

Sidebars: Throughout the text, you will find dedicated sections—some spanning a full page—that expand on specific technical processes, historical context, or complex physics.

Navigating the Sidebars

Sidebars are visually set apart from the main text and are numbered using a Chapter and Letter system (e.g., Sidebar 3A refers to the first sidebar in Chapter 3).

When you see a reference such as (see Sidebar 3A), it is a cue that more exhaustive technical data or a related visual deep-dive is available nearby. These are designed to be *modular*; you may read them as they appear to gain immediate context, or return to them later as a standalone technical resource.

# Chapter 1 - What is Color?

Color feels like a property of objects. A red apple appears red whether you are looking at it or not. In printing, that assumption causes confusion because it is only partly true. Objects do not *contain* color in the way they contain mass or temperature. Color is a sensation created by the visual system in response to light.

A working definition for this book is simple: color is what your brain experiences when your eyes respond to the spectrum of light entering them. The practical implication is equally simple: when the light entering your eye changes, the color experience can change, even if the object has not changed. In production environments, this is the root of many disagreements: proof versus press, plant versus plant, instrument versus visual, or *it matched yesterday* versus *it doesn't match today*.

---

## Note on Terminology

This opening chapter introduces several terms—such as spectrum, cones, metamerism, chromatic adaptation, viewing conditions, and measurement geometry—before they are fully developed. That is intentional. For now, treat them as labels for the main forces that shape color perception in print. Each will be defined precisely and expanded in later chapters, with practical examples and workflow guidance.

---

## Light is Energy

In the 17th century, Isaac Newton demonstrated that white light is not pure. By passing sunlight through a prism, he showed that color is present in the light (see Sidebar 1A).

Because light is energy, it can be described by its wavelength. Only a narrow band of wavelengths is visible to humans (see Sidebar 1B). Within that band, different wavelength ranges produce different color sensations. Shorter wavelengths are typically experienced as blue-violet, mid-range wavelengths as green, and longer

## Isaac Newton and His Prism Experiments

In the late 1600s, Isaac Newton demonstrated that color is a property of light itself, not a modification imposed by glass or other materials. In a darkened room, he passed a narrow beam of sunlight through a prism. The emerging light spread into a band of colors—a spectrum.

To test whether the prism created the colors, Newton directed one of the separated colors through a second prism. The color did not change. He then recombined the spectrum and recovered white light. These experiments showed that white light is composed of many wavelengths, each corresponding to a distinct color.

Newton divided the visible spectrum into seven named colors: red, orange, yellow, green, blue, indigo, and violet—remembered as "Roy G. Biv."

The number seven is not required by physics; the visible spectrum is continuous. Some historians believe Newton chose seven categories to parallel the seven notes of the musical scale, reflecting his interest in harmony and proportion.

Newton's work established a foundational principle of color science: color originates in light. Objects do not create color; they modify the light that strikes them.

**Sidebar 1A**

## The Visible Spectrum

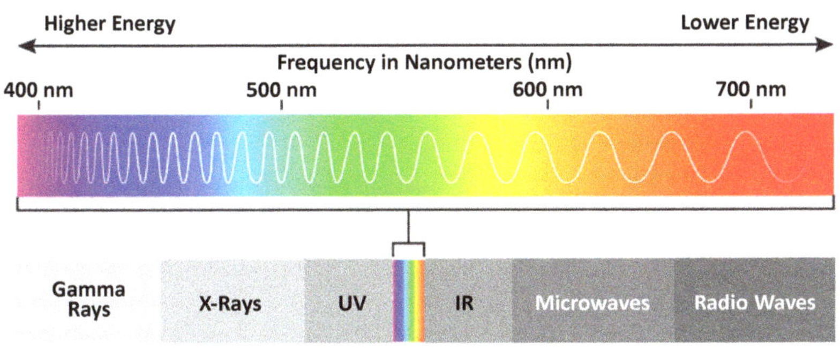


The electromagnetic spectrum is the complete range of electromagnetic radiation, organized by wavelength or energy. It includes Gamma Rays, X-Rays, Ultraviolet Radiation, Infrared Radiation, Microwaves, and Radio Waves. All of these are the same physical phenomenon—oscillating electric and magnetic fields—differing only in wavelength and energy. Longer wavelengths (such as radio waves) carry lower energy, while shorter wavelengths (such as X-rays and gamma rays) carry much higher energy. The electromagnetic spectrum spans an enormous range, far beyond what human senses can detect.

The visible spectrum is a very small portion of this larger spectrum, occupying only a narrow band between infrared and ultraviolet radiation. This is the range of wavelengths that the human visual system can detect, roughly from violet at the short-wavelength end to red at the long-wavelength end. Although physically small, the visible spectrum is critically important because it is the portion of electromagnetic radiation that enables human vision. All color perception—whether from light sources, printed materials, or objects in the world—is based entirely on how energy within this narrow band is emitted, reflected, or absorbed.

**Sidebar 1B**

wavelengths as yellow-orange-red. However, the eye does not measure wavelength directly like a spectrometer. Instead, it uses biological sensors that respond broadly to overlapping ranges of wavelengths.

The retina contains light-sensitive cells. Rods support vision in very low light, but they do not provide useful color information. Cones are the primary sensors for color vision (see Sidebar 1C). Most people have three types of cones with different spectral sensitivities. One is more sensitive to longer wavelengths, one peaks in the middle range, and one responds more strongly to shorter wavelengths. When light enters the eye, each cone type responds according to how much energy it receives in the wavelengths it is sensitive to. The brain interprets the relative pattern of responses across the cones as a color sensation.

These cone classes are properly described as long-, medium-, and short-wavelength sensitive. In more casual discussion, they are frequently shortened to red, green, and blue for compactness, even though cone sensitivities are broad and overlapping and do not correspond to single "pure" colors.

This point matters: there is no one-to-one *wavelength equals color* mapping. Many different spectral distributions can produce the same pattern of cone responses. When that happens, two samples can appear to be the same color even though their measured spectra are different. This phenomenon is called metamerism. Metamerism is not an exception in printing; it is routine. It is why a match under one light can change under another, and why *the numbers match* can still produce a visual disagreement when viewing conditions change.

The brain also does not report *raw sensor data*. It continuously adapts to the environment. It estimates the color of the illumination and partially *discounts* it so objects appear more stable than they otherwise would. This is called chromatic adaptation, and it is why a white sheet of paper appears white indoors under warm light and outdoors under cool daylight, even though the spectrum reaching your eye is very different in each case. Adaptation is helpful for

## The Human Eye

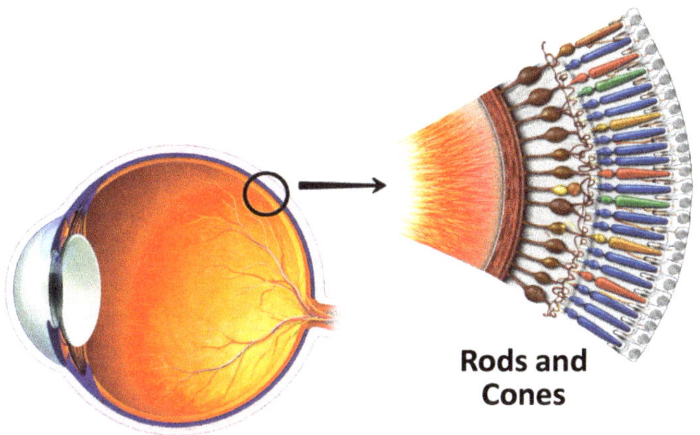


Cones are larger than rods, but they are far fewer in number. The human retina contains approximately 120 million rods and about 6 million cones. Cones are concentrated near the center of the retina in a region called the fovea. In this area, cones are densely packed and responsible for fine detail and high-resolution vision.

Cones also enable color perception. There are three types, sensitive to long (red), medium (green), and short (blue) wavelengths of light. Cones are less sensitive to light than rods, so they function best under bright conditions. In daylight, cones provide sharp, vivid color vision. As light levels decrease, cone activity declines, colors lose saturation, and vision becomes increasingly dependent on rods. In very low light, vision is essentially monochromatic.

**Sidebar 1C**

living in the world, but it complicates controlled comparison. If two people compare prints after looking at different surroundings, or in different rooms, or even at different times of day, their visual systems may be adapted differently before they ever look at the samples.

Color perception is also strongly influenced by context. Surrounding colors influence what a color appears to be. A neutral gray can look bluish on a warm background and yellowish on a cool background. Small shifts in brightness can change perceived chroma, and gloss or texture can change perceived lightness. For printing, this means that color is never just *the ink*. It is the ink, the substrate, the surface, the lighting, the viewing geometry, and the observer's visual state, all combined.

These realities do not make color unmanageable. They explain why managing color requires more than hitting a numeric target. If color is a perception created by the brain, then control depends on controlling the inputs to perception. In practical terms, the inputs you can control are the viewing conditions, the samples, and the measurement conditions you use to communicate and verify color.

This is why professional workflows treat lighting and viewing as part of the specification, not an afterthought. They define standard illuminants such as D50 for graphic arts viewing, account for optical brighteners in substrates, and insist that evaluations be made in a viewing booth with controlled illumination and a neutral surround. These controls do not eliminate subjectivity, but they dramatically reduce variability in what observers are being asked to judge.

This is also why instruments matter. A spectrophotometer does not *see* color the way you do. It measures reflected or transmitted light and reports numbers derived from standardized observer models. Those models exist precisely because color is a human sensation; they are a formal way to predict typical visual response under defined conditions. Instruments are essential because they provide repeatability and a shared language. But they only correlate well with visual judgment when the workflow controls the conditions the instrument assumptions depend on: illumination, geometry, sample presentation, and consistency of measurement mode.

If Chapter 1 has one message, it is this: color is not a thing you can hold. It is an experience your visual system produces. Printing succeeds when we stop treating color as a simple property of ink and start treating it as an engineered outcome of light, materials, and human perception.

---

## Key Takeaways

- Color is a sensation created by the brain in response to light; objects reflect spectra, they do not *contain* color.
- Human color vision depends on the relative responses of three cone types with broad, overlapping sensitivities.
- Different spectra can look the same under a given light, which is common in printing.
- The visual system adapts to illumination and context, changing perception even when samples do not change.
- Predictable color requires control of viewing conditions and alignment between visual evaluation and measurement.

---

## Practical Application

If a proof and a press sheet match under D50 in a viewing booth but disagree under warehouse LEDs, the workflow did not *fail*. You are seeing predictable consequences of different illumination and adaptation. The solution is not debate; the solution is specifying and using the correct viewing condition for approval.

# Chapter 2 - Light and Spectra

## Reflectance, Absorption, Fluorescence, and Why Spectra Matter

In Chapter 1, color was defined as a visual experience created by the brain in response to light. This chapter moves one step upstream. Before the eye and brain can respond, light must interact with materials. In printing, that interaction is where predictability is either built or lost.

Ink, paper, film, foil, coatings, and adhesives do not create color on their own. They modify the light that strikes them. To manage color reliably, it is not enough to describe materials as *red ink* or *white paper*. What matters is how those materials change the distribution of light energy across wavelengths—and how consistently they do so.

A spectrum is a description of how much light exists at each wavelength across the visible range. It shows the distribution of energy from shorter to longer wavelengths rather than assigning a color name. Each light source has its own spectrum. Each material also has a spectral behavior, describing how it absorbs and reflects different wavelengths. The plural of spectrum is *spectra*.

The spectrum that ultimately reaches the observer is the result of the light source interacting with the material. Human vision responds directly to that resulting spectral distribution through the combined activity of the cone receptors, producing the perception of color. A spectrophotometer measures the reflected spectrum and calculates color values using a defined *illuminant* and *standard observer*.

This chapter introduces spectra as the common language that connects illumination, materials, visual appearance, and measurement.

## The Spectrum as a Description of Light

Light can be described by its spectral power distribution: how much energy exists at each wavelength. Any real light source—daylight, fluorescent, LED, tungsten—has its own spectral shape. Two lights can appear similarly "white" to the eye while having very different spectral distributions (see Sidebar 2A).

The same idea applies after light interacts with an object. When light strikes a printed surface, some wavelengths are absorbed, some are reflected, and some may be re-emitted. The result is a reflected spectrum. That reflected spectrum is the physical signal that both the eye and a spectrophotometer receive.

From a workflow perspective, this matters because spectra are objective. They exist whether or not someone is looking at them. Visual color judgments and colorimetric numbers are both interpretations of spectra under defined assumptions.

## Reflectance and Absorption

Most printing materials are reflective. Light hits the surface, interacts with inks and the substrate, and a portion of the incident light is reflected back toward the viewer.

Printing Inks work primarily by absorption. Cyan, magenta, and yellow inks selectively absorb parts of the visible spectrum and reflect the rest. Where inks overlap, absorption increases. The reflected spectrum becomes more sculpted as more ink is added.

Substrates also contribute. Paper is not a neutral reflector. Different papers reflect different amounts of light at different wavelengths, especially in the blue region where optical brighteners may be present. Films, foils, and metallized materials introduce additional complexity through surface reflectance and specular components.

The key point is that the reflected spectrum is the combined result of illumination, ink absorption, and substrate reflectance. Changing

## Power Distribution Curves

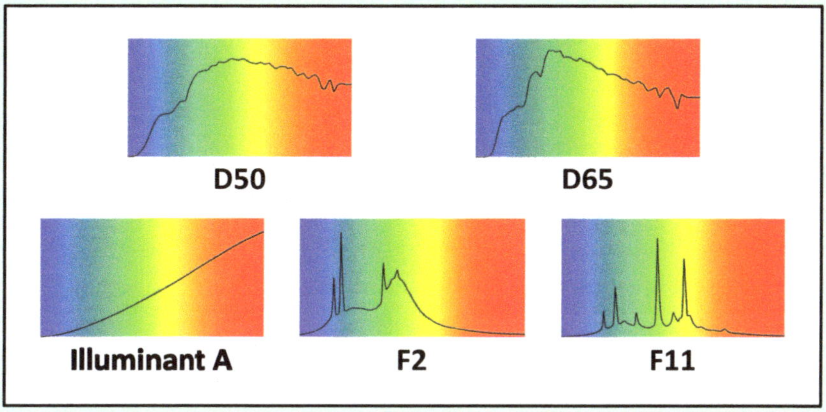


Power distribution curves describe how much radiant energy a light source emits at each wavelength across the visible spectrum. In structure, they resemble spectral reflectance curves: both are plotted as relative values versus wavelength. The key difference is what they represent. A reflectance curve shows how a material reflects incident light. A power distribution curve shows how a light source produces light.

For illuminants such as D50, D65, Illuminant A, F2, and F11, the curve reveals the character of the source. Daylight illuminants (D50, D65) exhibit broad, relatively smooth energy distributions. Illuminant A shows a continuous rise toward longer wavelengths, consistent with a tungsten source. Fluorescent sources such as F2 and F11 display sharp peaks at specific wavelengths rather than a smooth distribution of energy. These peaks directly influence color appearance, measurement results, and metamerism.

**Sidebar 2A**

any one of those changes the spectrum, even if the printed image looks *the same* at a glance.

---

## Why *Color Names* Are Not Enough

Describing materials as *warm white*, *bright yellow*, or *deep red* is convenient but incomplete. These labels collapse complex spectral behavior into a single impression. Two materials with the same color name can have very different spectra.

This is the root of many production surprises. A proof and a press sheet can look similar under one light and diverge under another because their spectra differ, even though their visual appearance happened to align under a specific condition. Chapter 1 introduced this as metamerism. Spectra explain why it occurs.

When workflows rely only on visual matches or numeric targets without regard to spectral behavior, they are vulnerable to illumination changes, substrate substitutions, and supplier variation.

---

## Fluorescence and Optical Brighteners

Some materials do more than reflect light. They fluoresce. Fluorescence occurs when a material absorbs light at one wavelength (typically in the ultraviolet region) and re-emits it at a longer wavelength within the visible spectrum (see Sidebar 2B).

Optical brightening agents (OBAs) in paper are the most common example in printing. OBAs absorb energy in the near-UV range and emit blue light. This increases apparent brightness and shifts the paper's appearance toward a cooler white.

Fluorescence complicates both visual evaluation and measurement. The amount of fluorescence depends on the illumination. A light source with more UV content will excite OBAs more strongly than one with little or no UV. As a result, the same paper can appear different under different light sources, even if the inks are unchanged.

## Spectral Reflectance of OBA Brightened Paper

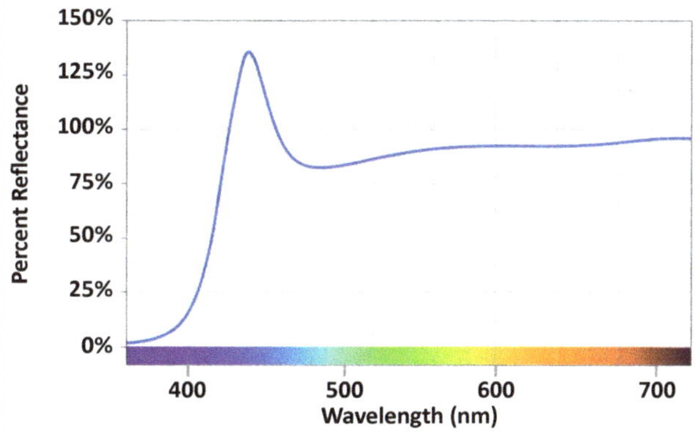


Some paper and board grades contain optical brightening agents (OBAs). These materials absorb ultraviolet energy—primarily in the near-UV region—and re-emit it as visible blue-violet light. This process is fluorescence.

Because a spectrophotometer reports reflectance relative to the incident light at each wavelength, the added fluorescent emission can cause measured reflectance values to exceed 100% in the violet region. The material is not *more reflective* than a perfect mirror. Instead, it is emitting additional visible energy after absorbing ultraviolet radiation.

The magnitude of this effect depends on the amount of OBAs present and the UV content of the illuminant. Under UV-rich light sources, the violet peak becomes more pronounced. Under UV-filtered conditions, the peak is reduced.

For this reason, measurement mode (M0, M1, M2, or M3) and UV control directly influence results on OBA-containing substrates. Fluorescence must be managed intentionally to achieve consistent and meaningful color data.

**Sidebar 2B**

From a measurement standpoint, fluorescence breaks simple assumptions. A spectrophotometer user must decide whether to include, exclude, or control UV energy in its illumination. Later chapters will address measurement modes in detail. For now, the important concept is that fluorescence makes spectra illumination-dependent in a way that simple reflectance does not.

---

## Spectra as the Bridge Between Vision and Measurement

The eye does not see spectra directly. It integrates them through cone responses. A spectrophotometer does not "see color" at all. It measures spectral data and applies standardized observer models to *predict* visual response.

Despite these differences, spectra are the common ground. If two samples have identical reflected spectra, they will match visually under any illumination that does not introduce fluorescence effects. If their spectra differ, visual matches are conditional.

This is why spectra matter even for people who never look at a spectral curve. Spectral differences explain why matches break, why approvals fail under new lighting, and why instrument agreement does not always guarantee visual agreement.

Understanding spectra does not mean memorizing curves. It means recognizing that color numbers are summaries of spectral data, not replacements for it.

---

## Practical Consequences in Print Workflows

Several common situations trace directly back to spectral behavior:

- A spot color matches on coated and uncoated drawdowns but shifts on press because the substrate reflectance differs.

- A brand color approved under D50 looks dull under retail LEDs because the illumination spectrum is different.
- Two inks measure the same L*a*b* values under one condition but diverge visually after a supplier change.
- A paper substitution introduces unexpected brightness changes due to different OBA levels.

In each case, the underlying issue is not opinion or instrument error. It is physics. The spectra reaching the eye changed.

---

## Controlling What You Can Control

You cannot control how every observer's visual system works. You can control illumination, substrates, inks, viewing conditions, and measurement assumptions.

Professional color workflows do not eliminate spectral differences. They manage them by standardizing conditions and aligning expectations. Viewing standards, measurement modes, and material specifications exist precisely to limit how much spectral variability is allowed to influence decisions.

Chapter 3 will return to the human side—adaptation, surround, and context effects. Later chapters will show how measurement standards formalize these ideas. For now, the foundation is simple: spectra are the physical reality behind every color judgment and every color number.

---

## Key Takeaways

- Light interacting with materials produces a reflected spectrum, which is the true physical signal behind color.
- Inks primarily shape spectra through absorption; substrates and coatings contribute their own spectral behavior.

- Different spectra can produce similar visual color under one condition but diverge under another.
- Fluorescence, especially from optical brighteners, makes color strongly dependent on illumination.
- Spectra are the bridge between visual perception and instrumental measurement.

---

## Practical Application

If a color match fails when lighting or paper changes, do not start by blaming the press or the instrument. Start by asking what changed in the spectrum reaching the eye. In most cases, the answer will point directly to the root cause and the appropriate control.

# Chapter 3 - Human Vision in Production

## Adaptation, Surround, Contrast, and Metamerism

By this point, two facts are established. First, color is a perceptual experience created by the visual system. Second, that experience is driven by the spectrum of light reaching the eye. What remains is the most uncomfortable part for production environments: the human visual system is not a passive sensor. It is adaptive, contextual, and optimized for survival, not for press approval.

This chapter explains the specific behaviors of human vision that most often disrupt color decisions in printing. These behaviors are not flaws. They are normal, predictable responses of a biological system. Problems arise only when workflows ignore them.

---

### Visual Adaptation Is Continuous

The visual system is always adapting. It adjusts its sensitivity based on the lighting environment and recent visual history. This process begins within seconds and continues over minutes.

Chromatic adaptation is the most relevant form for printing. The brain estimates the color of the illumination and partially discounts it so objects appear more stable. This is why a white page appears white under warm indoor lighting and cool daylight, even though the spectra reaching the eye are very different.

In production, adaptation becomes a variable whenever observers enter a viewing environment from elsewhere. Someone walking in from a warehouse, an office, or outdoors brings a different state of adaptation than someone who has been standing in a viewing booth for several minutes. If evaluations begin immediately, judgments will differ even if the samples do not.

This is not subjective behavior. It is physiology.

## Surround and Context Shape Perception

Color perception depends strongly on what surrounds the sample. The visual system interprets colors relative to their context, not in isolation.

A neutral gray placed on a dark background will appear lighter than the same gray on a light background. A saturated color surrounded by complementary hues will appear more vivid than it does in isolation. These effects occur automatically and unconsciously.

In printing, surround effects show up in predictable ways:

- Proofs evaluated on a bright white desk look different than the same proofs viewed in a neutral booth.
- Press sheets judged while still surrounded by unprinted substrate appear different once trimmed.
- Small samples compared against large, saturated backgrounds shift in perceived chroma.
- When the same brand logo is printed in the same unit on a mixed form, but surrounded by different background colors, it will appear to change even though the ink, substrate, and press conditions are all identical.

This is why professional viewing environments use neutral gray surrounds and controlled sample presentation. The goal is not to eliminate context effects—that is impossible—but to make them consistent.

## Simultaneous Contrast Effects Are Local and Immediate

Simultaneous contrast occurs when adjacent colors influence each other's appearance. A color can appear darker, lighter, more saturated, or shifted in hue depending on neighboring colors.

This effect is strongest at boundaries and edges. In practical terms, a spot color surrounded by dark imagery may appear lighter than the same color surrounded by white space, even though the ink film and measurement are identical.

Simultaneous contrast effects explain why isolated color swatches often behave differently than the same colors embedded in final artwork. They also explain why approving an ink drawdown does not always guarantee satisfaction in the finished design.

Simultaneous contrast is not an error condition. It is how vision extracts information. Workflows fail when they assume patch behavior predicts appearance in all contexts.

---

## Metamerism Revisited: A Human Problem, Not an Instrument Problem

Chapter 1 introduced metamerism as the condition where different spectra produce the same color appearance under one light source but not another. Chapter 2 showed how spectral differences make this possible. This chapter closes the loop.

Metamerism exists because the visual system compresses spectral information into three cone responses. That compression is useful but lossy. Information is discarded. When illumination changes, the discarded information can no longer be ignored.

This is why metamerism cannot be "fixed" by better eyesight or stronger opinions. It is inherent in trichromatic vision. Instruments reveal metamerism because they retain spectral detail. Visual evaluation reveals it when conditions change.

Understanding this reframes many disputes. A disagreement caused by metamerism is not a failure of quality control or skill. It is a predictable outcome of how humans see.

---

## Observer Variability Is Real, but Bounded

No two observers have identical visual systems. Cone sensitivities vary slightly from person to person. With age, the eye's lens gradually yellows and absorbs more short-wavelength (blue) light.

As blue light transmission decreases, perception can shift subtly toward yellow. Fatigue, health, and prior visual adaptation also influence perception.

However, these differences are smaller than many people assume. Large disagreements are rarely caused by innate observer differences alone. They are usually amplified by uncontrolled adaptation, surround, lighting, or sample presentation.

Standard observers (see Sidebars 3A & 3B) exist for a reason. While no standard matches every individual perfectly, they provide a common reference that works well when viewing conditions are controlled. Instruments and standards are not meant to replace vision. They are meant to stabilize it.

---

## Why *Looks Good to Me* Is Not a Workflow

Visual evaluation is essential in printing. Final approval is always visual. But vision without controls is unreliable as a decision system.

When workflows rely on unconditioned visual judgment, outcomes depend on who is present, where they are standing, what they looked at five minutes ago, and what surrounds the sample. These are not acceptable variables in production.

Professional workflows treat the observer as part of the system. They define adaptation time, surround color, viewing geometry, illumination, and sample handling. These controls do not remove judgment. They make judgment repeatable.

---

## Designing for Human Vision Instead of Fighting It

Successful color programs do not try to out-argue human perception. They design around it.

- They specify viewing conditions so adaptation is predictable.

- They control surrounds so context effects are consistent.
- They understand where contrast will change appearance and plan approvals accordingly.
- They recognize metamerism as a risk and manage it through material choices and lighting standards.

When these elements are aligned, visual evaluation becomes a strength instead of a liability.

---

## Visual Perception Demonstrations

Understanding L*a*b*, spectral data, and observer models is only part of the story. Human perception is not uniform. Millions of people experience measurable differences in color discrimination, and those differences directly affect design, branding, packaging, and product usability.

Our companion website provides structured video demonstrations of common visual perception issues, such as, retinal fatigue, simultaneous contrast, and color memory.

The goal is practical understanding. These demonstrations are designed to help you:

- Anticipate perceptual failures before production
- Understand the limitations of human color vision
- Improve compliance and customer satisfaction

If you work in production, branding, product design, or color quality control, this material will sharpen your ability to design for the full range of human observers—not just the average one. Please visit:

**rgbinder.com/colortheorybook**

## 2° Standard Observer

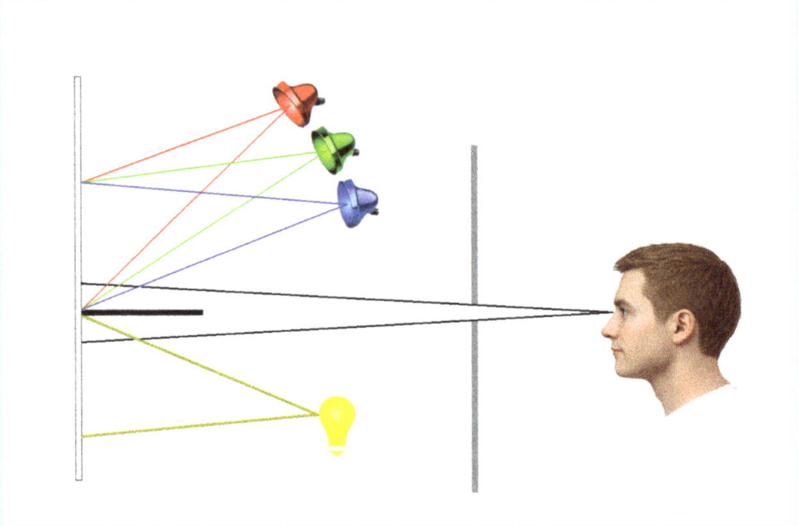

In the 1920s, the Commission Internationale de l'Éclairage (CIE) conducted controlled visual experiments to quantify human color matching. Observers viewed a small circular test field, approximately 2 degrees of visual angle—about the size of a thumbnail held at arm's length.

Participants adjusted the intensities of three primary lights (red, green, and blue) until the mixture visually matched a monochromatic test light. By repeating this process across wavelengths, researchers derived the color matching functions that describe average human cone response.

The result was the CIE 1931 2° Standard Observer. It represents foveal vision—color perception dominated by the central retina, where cone density is highest. Because most color comparison tasks involve small fields or fine detail, the 2° observer remains widely used for color measurement in printing.

**Sidebar 3A**

## 10° Standard Observer

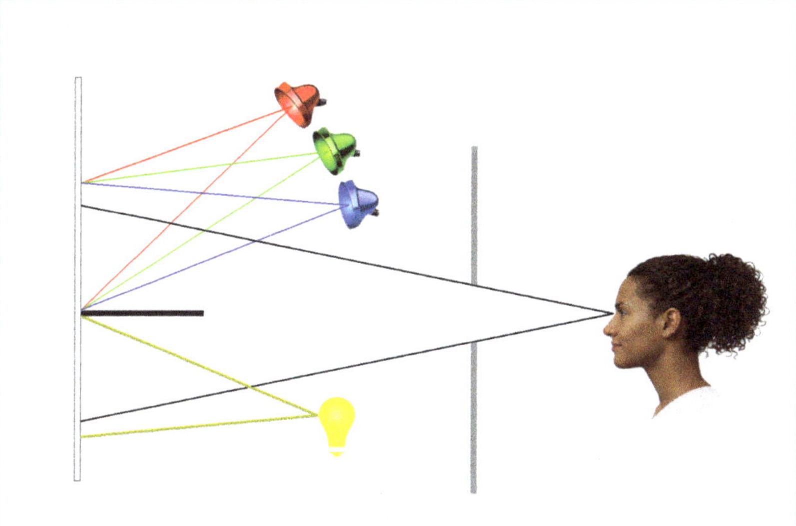

Subsequent research showed that color perception shifts slightly when larger areas are viewed. As the visual field expands beyond the central fovea, more of the retina contributes to the response, including areas with different cone distributions.

To account for this, the CIE conducted additional experiments using a larger, 10-degree field of view. The resulting CIE 1964 10° Standard Observer reflects average human response over a broader visual area.

Differences between the 2° and 10° observers are subtle but measurable, particularly in the blue region. For large uniform areas—such as textiles, plastics, or wide-format print—the 10° observer may better represent perceived color. For small patches and fine detail, the 2° observer is typically appropriate.

**Sidebar 3B**

## Key Takeaways

- The visual system is adaptive; judgments depend on recent visual history and lighting exposure.
- Surround and context strongly influence perceived color and must be controlled.
- Simultaneous contrast effects explain why isolated swatches and finished artwork can behave differently.
- Metamerism is a consequence of human vision, not an inspection failure.
- Reliable visual evaluation requires defined conditions, not stronger opinions.

## Practical Application

If two people disagree on a color match, do not ask who is right. Ask where they were standing, what they looked at before, what surrounds the sample, and how long they have been in the viewing environment. In most cases, correcting those conditions resolves the disagreement without changing the ink.

# Chapter 4 - Viewing and Illumination

## Standards, LEDs, OBAs, and Approval Discipline

By now, three realities are clear. Color is a perceptual experience. That experience is driven by spectra. And the human visual system adapts continuously to its environment. Taken together, these facts lead to an unavoidable conclusion: *visual evaluation without controlled viewing conditions is not reliable.*

This chapter explains how the printing industry imposes order on human vision. It introduces standard illuminants, viewing environments, and approval discipline—not as formalities, but as tools for reducing variability and avoiding predictable failure modes.

---

### Illumination Is Part of the Specification

In uncontrolled environments, lighting is treated as background. In professional color workflows, illumination is a defined variable.

The spectral power distribution of a physical light source determines the spectral stimulus reflected from the object and reaching the eye. If the illumination changes, the reflected spectrum changes, even when the object itself does not. Two prints may match under one condition and diverge under another because the illumination spectrum has changed.

Color standards therefore specify illuminants. An illuminant is not a physical lamp. It is a mathematically defined spectral power distribution used as a reference condition for measurement and calculation of tristimulus values.

In practice, viewing booths are engineered so that their physical light sources approximate these standardized illuminants as closely as possible. The *illuminant* defines the target condition; the *light source* is the physical device used to realize it.

## D50 and D65: Why They Exist

In graphic arts, D50 is the dominant standard viewing illuminant. It approximates mid-day daylight with a correlated color temperature of approximately 5000 K. It was selected to align with long-standing industry practice and with CIE/ISO colorimetric reference conditions used for measurement and ICC workflows.

D65, at approximately 6500 K, represents a cooler daylight condition. It is widely used in display technologies and in industries outside traditional print, and it is *sometimes* specified when print is intended to visually harmonize with on-screen content or broader daylight environments.

In addition to daylight illuminants, secondary illuminants are often specified for validation. Retail environments frequently use fluorescent or LED lighting with spiked spectral power distributions. For this reason, standards and brand specifications for packaging may include fluorescent illuminants such as F2 or F11 as secondary viewing conditions. These are used to evaluate metamerism risk and to simulate point-of-sale lighting.

The important point is not which illuminant is *better*. The requirement is that a shared reference illuminant be defined. Color decisions are only meaningful when all parties evaluate the work under the same specified spectral condition.

Mixing viewing conditions—approving under one light source and producing under assumptions tied to another—is a common root cause of disputes.

## LED Lighting and Spectral Mismatch

Modern LEDs have made lighting more efficient and more complex. Many LED sources achieve a desired appearance using narrow spectral peaks rather than smooth distributions.

Two LEDs with the same correlated color temperature can have very different spectra. As a result, they can produce very different color appearances for the same print, especially when OBAs or certain pigments are involved.

This is why *it's a 5000 K LED* is not sufficient. Spectral quality matters. Viewing standards now include additional metrics to constrain LED behavior so that they approximate traditional illuminants more reliably.

From a workflow standpoint, *uncontrolled* LED lighting is one of the fastest ways to reintroduce variability that standards were designed to eliminate.

Modern light booth manufacturers have addressed this issue directly. Contemporary LED viewing systems are not simple replacements for fluorescent tubes. They are engineered to approximate standardized illuminants such as D50 or D65 with tight spectral tolerances, controlled chromaticity, and verified conformance to ISO and industry specifications. The spectral design, phosphor blends, optical mixing, and calibration processes are deliberate.

It is not acceptable to replace fluorescent tubes with off-the-shelf LED lamps and assume equivalence. Even if the correlated color temperature appears correct, the underlying spectral power distribution may differ significantly. Because color evaluation depends on spectral structure—not just CCT—uncontrolled substitution can introduce metameric shifts, visual mismatch, and disagreement across the workflow.

---

## Optical Brighteners Change the Rules

Optical brightening agents (OBAs) complicate viewing because they rely on fluorescence. The amount of blue light emitted by OBAs depends on how much short-wavelength energy is present in the illumination.

A paper with OBAs can look brighter and cooler under a light rich in UV energy and duller under a light that suppresses it. This effect is not subtle. It can materially change approvals.

For this reason, viewing conditions must be chosen deliberately when OBAs are involved. If the approval light excites OBAs differently than the production or retail environment, disagreements are inevitable.

The workflow question is not: *Are OBAs good or bad?* The question is: *Under what illumination are we approving color, and does that illumination reflect how the product will be judged downstream?*

---

## Viewing Geometry and Booth Design

Illumination alone is not enough. How samples are viewed matters.

Viewing booths are designed to provide:

- Controlled, standardized illumination
- Neutral gray surrounds
- Isolation from ambient light

Neutral gray surrounds reduce context effects. Isolation prevents contamination from nearby lighting sources.

Using a booth does not guarantee agreement. Using a booth incorrectly guarantees disagreement. Samples held at odd angles, viewed against uncontrolled surroundings, or evaluated by viewers wearing brightly colored clothing undermine the purpose of the booth.

---

## Adaptation Time Is Not Optional

Human vision requires time to adapt to a new lighting environment. Immediate judgments are unreliable.

Professional practice assumes a short adaptation period before evaluation begins. This can be as simple as standing in the viewing environment and looking at a neutral reference for a minute or two before judging color.

Skipping adaptation is one of the most common and least acknowledged causes of disagreement. It costs nothing to fix and is rarely enforced.

---

## Approval Discipline Is a System, Not a Preference

Viewing conditions only matter if they are used consistently. This requires discipline.

Approval discipline includes:

- Defining the illumination under which approval is valid
- Evaluating samples only under that condition
- Avoiding side-by-side comparisons under mixed lighting
- Communicating clearly when approvals are conditional

An approval made under uncontrolled lighting is not "wrong." It is incomplete. If downstream evaluation occurs under a different condition, the approval no longer applies.

Professional workflows treat approvals as conditional statements: *approved under specified conditions*. When those conditions change, the approval must be reconsidered.

---

## Why This Matters Before Measurement

It may be tempting to jump directly to instruments and tolerances.

Measurement standards assume specific viewing conditions. If those assumptions are violated, numeric agreement does not guarantee visual agreement, and visual agreement may not correlate with numbers.

Viewing and illumination must be stabilized first. Measurement only makes sense once the visual reference is *defined*.

## Key Takeaways

- Illumination directly shapes color appearance and must be specified.
- D50, D65, and other standard illuminants exist to provide shared reference conditions, not to declare one *correct*.
- LED lighting introduces spectral variability that must be managed deliberately.
- Optical brighteners make color strongly dependent on illumination.
- Viewing booths, surrounds, and geometry reduce variability only when used correctly.
- Visual approvals are valid only under the conditions in which they were made.

## Practical Application

If a color is approved in a viewing booth under D50 but rejected under office lighting, the workflow has not failed. The approval conditions were violated. The correct response is to return to the defined viewing environment, not to renegotiate color on the fly.

# Chapter 5 - Measuring Color

## Instrument Fundamentals, Geometry, and Repeatability

At this point, the role of visual evaluation should be clear. Color decisions depend on light, materials, and the human visual system. Measurement enters the workflow not to replace vision, but to stabilize it.

Color instruments exist because visual judgment alone cannot deliver repeatability. People adapt. Lighting changes. Color memory is virtually nonexistent. Instruments provide a shared reference that does not depend on who is looking or when they are looking. Used correctly, they make color decisions data-driven rather than opinion-based. However, used incorrectly, they can create false certainty.

This chapter explains what color instruments actually measure, how they do it, and what must be controlled for measurements to be meaningful.

---

### What a Color Instrument Measures

A spectrophotometer does not measure *color*. It measures light.

Specifically, it measures the amount of light reflected[1] by a sample at discrete wavelength intervals across the visible spectrum. The result is spectral reflectance data that describes how the sample modifies the illumination. When plotted, it forms a reflectance curve.

---

[1] Depending on instrument design and application, a spectrophotometer may measure reflected, transmitted, or emitted radiant energy. Reflectance instruments are most common in printing and packaging workflows. Transmission instruments are used for materials such as films, plastics, and liquids. Emission measurements are used for light sources and displays.

Everything else—XYZ values, L*a*b* data, ΔE information—is derived from that spectral data using standardized observer models and defined viewing conditions.

This distinction matters. When someone says *the numbers match,* what they really mean is *the calculated values derived from measured spectra match under specific assumptions.* If those assumptions do not align with how the sample is viewed, agreement breaks down.

---

## Color Data Depends on Measurement Configuration

Every color measurement is defined by specific conditions:

- A selected illuminant
- A selected standard observer
- A defined geometry
- A defined measurement mode

These conditions are not optional. They are embedded in the reported result. If visual evaluation occurs under different conditions, disagreement *should be expected.*

This is why Chapters 3 and 4 precede this one. Measurement is only meaningful when the visual reference conditions are defined first.

---

## Measurement Geometry: How Light Interacts with the Sample

Measurement geometry describes the angular relationship between illumination and detection. It determines how surface effects such as gloss and texture influence the result.

Two geometries dominate printing:

- 45°/0° (or 0°/45°)
- d/8° (diffuse/8° also known as spherical)

A 45°/0° geometry illuminates the sample at a 45° angle and measures perpendicular to the surface. This geometry reduces the influence of surface gloss and more closely mimics how flat samples are typically viewed by human observers (see Sidebar 5A).

A d/8° geometry uses diffuse (hemispherical) illumination from an integrating sphere and measures the integrated reflected light along a defined 8° axis. The result represents an average reflectance over many angles. Depending on configuration, the specular component may be included (SCI) or excluded (SCE) (see Sidebar 5B).

Neither geometry is *more accurate* in isolation. Each answers a different question. Problems arise when geometry is changed without understanding the visual or workflow implications.

Geometry mismatch is occasionally a hidden cause of disagreement between sites or between instruments.

A 45°/0° geometry is appropriate for most conventional print applications. However, for highly reflective surfaces—such as transparent inks printed on metallic foil—or for textured substrates where directional illumination can produce unstable readings, a d/8° geometry is required to obtain stable and representative measurements.

---

## Process Control Depends on Repeatability

In production environments, both accuracy and repeatability are essential. Accuracy ensures that measurements correspond to defined visual and instrumental conditions. Repeatability ensures that results remain consistent under those same conditions.

An instrument that produces stable, consistent results can be used to monitor and control process variation. Even small inconsistencies in measurement technique can obscure real process changes or create the appearance of variation where none exists.

This is why procedure discipline matters when measuring samples:

- Consistent positioning
- Consistent backing
- Clean optics
- Stable sample handling
- Consistent ambient temperature

Small deviations in technique can produce statistically significant differences.

Repeatability is not a property of the instrument alone. It is a property of the instrument, the operator, and the measurement procedure working together.

## 45°/0° Geometry

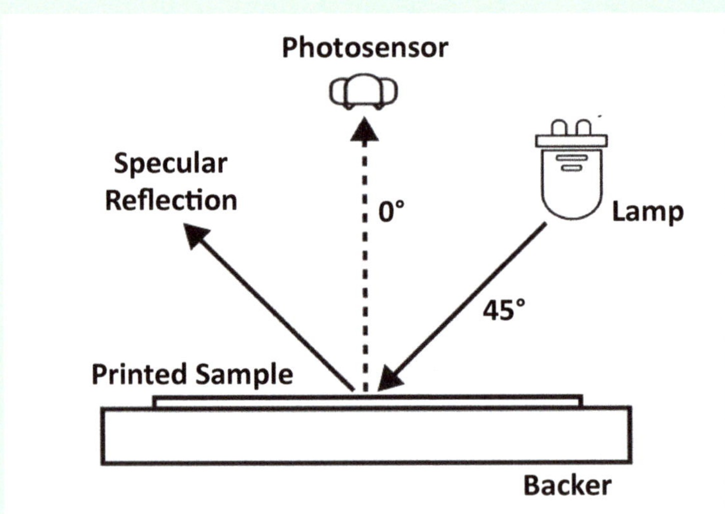


**Sidebar 5A**

## d/8° Geometry

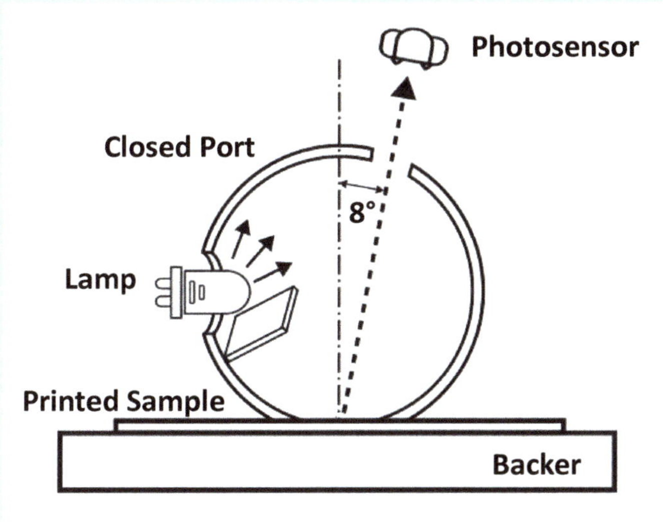


Specular Included (SCI)

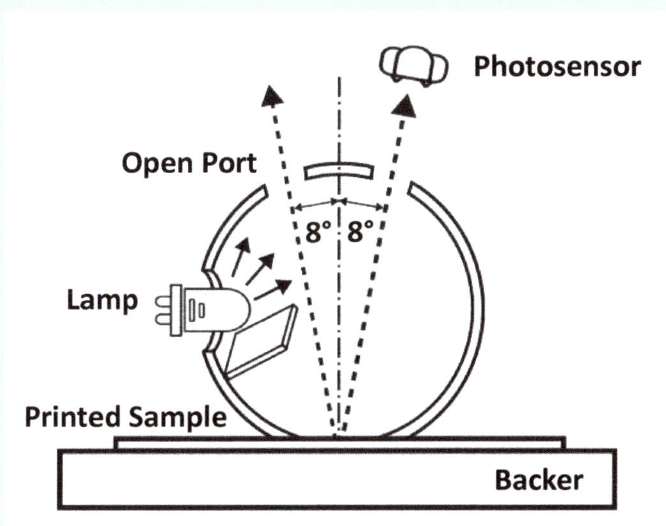


Specular Excluded (SCE)

**Sidebar 5B**

## Aperture Size and Sample Structure

Aperture size determines how much of the sample is averaged in a single measurement (see Sidebar 5C).

Small apertures are sensitive to local variation. Large apertures average over more structure. In printing, dot structure, screen patterns, and substrate texture mean that aperture choice can influence results.

There is no universally correct aperture. The correct choice depends on what you are trying to control: local defects, overall appearance, or compliance with a standard.

Changing aperture size without adjusting expectations is another common source of confusion.

## Why Instruments Disagree

When two instruments disagree, the cause is almost never mysterious. Typical reasons include:

- Different geometry
- Different measurement modes
- Different aperture sizes
- Different calibration states
- Different sample handling

Less commonly, instruments may be out of tolerance or damaged.

The correct response to disagreement is not to choose the instrument you prefer. It is to identify which assumptions differ and decide which configuration aligns with the workflow.

## Aperture Size

Aperture size determines how much surface area is included in a single measurement. In press environments, there is often pressure to use the smallest possible aperture to conserve space on the press sheet. While small apertures are practical, they must still be large enough to produce representative and statistically stable readings.

When measuring halftone patches for dot area or tone value, the aperture must be large enough to sample a sufficient number of halftone dots to avoid sampling bias. The required size depends on the line screen (or screening method) in use. The measurement area should include many individual dots so the reading represents the intended tone value rather than isolated dot structure.

Larger apertures are also required when measuring samples that exhibit spatial variation. Examples include ink drawdowns with uneven film thickness, textured or absorbent substrates, and corrugated materials where flute structure can influence reflectance. In these cases, a larger aperture averages local variation and produces more stable, representative data.

Aperture selection is therefore a balance between conserving space for target patches and ensuring measurement integrity. The smallest aperture is not automatically the best choice; it must be appropriate for the structure and variability of the sample being measured.

**Sidebar 5C**

## Measurement Is a Tool, Not a Verdict

Color instruments are powerful because they are impartial. They do not tire or adapt to surroundings. They report exactly what they are configured to report.

An instrument cannot determine whether a color is acceptable. It reports numerical differences—such as ΔE—relative to a defined reference under defined conditions. The acceptance tolerance is established by people, specifications, contracts, or brand standards. The instrument does not set the tolerance; it only evaluates against it.

If a measurement falls within tolerance but the visual match is unacceptable, the result should not be ignored or blindly accepted. Instead, the discrepancy must be investigated. Possible causes include incorrect measurement configuration, mismatched viewing conditions, gloss or texture effects, fluorescence, metamerism, or an inappropriate tolerance for the application.

Successful workflows treat instruments as decision-support tools. They provide objective data, but acceptance remains a human responsibility informed by both measurement and visual evaluation.

This keeps the authority of measurement intact while clearly assigning accountability to the defined standard and the people who set it.

## Preparing for Measurement Standards

This chapter has intentionally avoided deep discussion of measurement modes, polarization, or illuminant control. Those topics formalize the assumptions introduced here and will be addressed next.

For now, the takeaway is simple: *measurement only works when everyone agrees on what is being measured and why.*

## Key Takeaways

- Color instruments measure reflected or transmitted light, not color itself.
- All reported color values are derived from spectral data under defined assumptions.
- Measurement geometry strongly influences how surface effects are treated.
- Repeatability and disciplined procedure are essential for effective process control in production.
- Instrument disagreement usually traces to configuration differences, not instrument failure.
- Measurement supports visual decisions; it does not replace them.

## Practical Application

If two sites report different numbers for the same sample, do not ask which site is wrong. Ask which geometry, aperture, and conditions are being used and whether they match the approved viewing conditions. Alignment resolves most disputes without recalibrating color targets.

# Chapter 6 - Measurement Standards

## Modes, Geometry Variants, and When Mismatch Breaks Workflows

Chapters 4 and 5 established two prerequisites for reliable color control: defined viewing conditions and disciplined measurement procedures. This chapter formalizes those ideas. Measurement standards exist to lock assumptions in place so numbers remain comparable across time, people, and locations.

Most color workflow failures involving instruments do not come from bad targets or poor calibration. They come from *mismatched measurement assumptions*. This chapter explains the most important standards used in printing, what problem each one solves, and what happens when they are mixed without intent.

---

### Why Measurement Standards Exist

A spectrophotometer measures spectral data, but the way it illuminates the sample and interprets the result must be standardized. Without standards, two instruments measuring the same sample could legitimately report different results, even if both are functioning correctly.

Measurement standards define:

- How the sample is illuminated
- How results are intended to correlate with visual evaluation
- How fluorescence is handled
- Whether surface reflections are included

Standards do not guarantee visual agreement. They guarantee *comparability*. That distinction matters.

## Measurement Modes and Fluorescence Control

Modern print workflows must account for fluorescence, particularly from optical brightening agents. Measurement modes exist in part to control how short-wavelength energy is handled during measurement (see Sidebar 6A).

Commonly used modes differ in whether ultraviolet energy is included, excluded, or tightly controlled in the instrument's illumination. The intent is to manage how OBAs contribute to measured results.

The critical rule is simple: *the mode used to create a reference must match the mode used to verify it.* If the mode changes, the numbers are no longer comparable, even if the sample has not changed.

Mode mismatch is one cause of unexplained color measurement variation in multi-site programs.

## Specular Inclusion and Exclusion

In d/8° measurement geometry, the instrument allows control over whether the specular component—the mirror-like reflection from the surface—is included or excluded from the measurement (see Sidebar 5B).

**Specular Component Included (SCI)** captures total reflected light. This includes both diffuse reflectance from the colorant and specular reflectance from the surface. Because the specular component is included in the measurement geometry, surface texture differences (smooth versus textured) tend to produce consistent color readings. For this reason, SCI is often used to represent the intrinsic or "true" color of a material, independent of surface finish.

**Specular Component Excluded (SCE)** removes the mirror-like reflection and measures primarily the diffuse reflectance. As a result, differences in surface gloss or texture influence the reported values. Smooth areas may measure darker than textured areas,

# 45°/0° Measurement Modes

**M0: The Legacy Condition**

M0 includes all historical data from the time before these modes were defined and introduced (2009). At that time, devices typically used unfiltered tungsten light. It remains dominant because most long-established brand and non-brand color libraries were created in M0. However, it can lack consistency between measurements taken with instruments from different suppliers when measuring samples on substrates with optical brighteners.

**M1: Designed for Graphic Arts**

M1 simulates true D50 more accurately and gives consistent results across devices, on substrates containing OBAs. It is now the recommended and forward-looking standard for printing.

Brand programs transitioning to M1 must manage workflows carefully to avoid mixing legacy (M0-based) and modern (M1-based) standards.

**M2: UV-Cut**

M2 intentionally removes UV to eliminate fluorescence, which is useful when building ICC profiles. It is typically not used for day-to-day press control.

**M3: Polarized Measurements**

Used primarily to suppress specular reflection on glossy or textured surfaces. M3 can help in specific diagnostic scenarios but produces values that do not correlate well with visual appearance. Printing workflows rarely rely on M3.

**Sidebar 6A**

mirroring how human observers perceive them under directional lighting. SCE therefore tends to correlate more closely with visual appearance.

Neither mode is inherently correct. SCI is preferred when the objective is to measure material color independent of surface effects. SCE is preferred when the objective is to capture appearance as perceived, including the influence of gloss and texture.

What is non-negotiable is consistency. A standard measured with SCI cannot be meaningfully compared to samples measured with SCE. The geometry and specular mode are part of the measurement specification and must match.

---

## Geometry Variants and Workflow Intent

Geometry determines how illumination interacts with the sample surface. As discussed in Chapter 5, different geometries answer different questions.

Measurement standards formalize geometry choices so that:

- Results are comparable across instruments
- Expectations align with visual evaluation conditions
- Surface effects are treated consistently

Changing geometry mid-workflow is equivalent to changing viewing conditions mid-approval. It invalidates comparisons.

---

## Polarization

Some measurement configurations may use polarization to reduce surface reflection effects.

Polarized measurement is uncommon in North American colorimetric workflows. It is more commonly used in European density-based process control, where minimizing surface reflections may improve density measurement stability.

---

## Calibration Ensures Accuracy

Calibration aligns an instrument to a reference.

However, two instruments can be perfectly calibrated and still disagree if they use different modes, geometry assumptions, or sample handling procedures. Calibration ensures the instrument is behaving as designed. Standards ensure that design aligns with workflow intent.

---

## When Mismatch Becomes Invisible

One of the most dangerous aspects of standards mismatch is that it is not always obvious. Numbers may appear reasonable. Trends may look stable. Disagreements may surface only when lighting changes, materials shift, or samples move between sites.

By the time mismatch is noticed, it is often blamed on production rather than configuration.

This is why standards must be documented and enforced, not assumed.

---

## Designing a Measurement Standard Set

A complete measurement specification includes:

- Geometry
- Measurement mode (if 45°/0°)
- Specular handling (if d/8°)
- Polarization (if used)
- Aperture selection

- Backing material
- In some cases instrument manufacturer and model may be specified

These elements should be defined once and applied everywhere. Exceptions should be rare and documented.

---

## Key Takeaways

- Measurement standards exist to guarantee comparability, not visual perfection.
- Instrument modes must match between reference and verification.
- Geometry changes invalidate comparisons.
- Specular inclusion and exclusion change what is being measured.
- Calibration alone does not ensure agreement.
- Many measurement disputes trace back to mismatched assumptions, not bad color.

---

## Practical Application

If a color appears to drift numerically while remaining visually stable, check measurement modes and geometry before adjusting targets or tolerances. The "drift" may be an artifact of mismatch, not a real change in production.

# Chapter 7 - From Spectra to L*a*b*
# How Instruments Turn Light into Numbers

Up to this point, we have discussed instruments measuring spectral reflectance under defined conditions. The result is a reflectance curve describing how the material reflects light at discrete wavelength intervals across the visible spectrum. To be useful in production, that spectral data must be converted into colorimetric values.

This chapter explains how spectral reflectance is integrated to produce XYZ, and how XYZ is subsequently transformed into L*a*b* and L*C*h°.[1] These are not additional measurements. They are standardized mathematical transformations that estimate how a defined human observer would respond under a specified illuminant.

---

## From Spectral Reflectance to XYZ

A spectrophotometer measures spectral reflectance by sampling the spectrum at discrete wavelength intervals. That curve does not inherently define color; it quantifies how the object reflects incident radiation. In printing, we alter the substrate's spectral behavior to control the light returned to the observer. To assign color meaning to that data, the reflectance must be transformed through a specified illuminant and observer model.

To calculate XYZ values, three elements are combined:

- The measured spectral reflectance of the sample
- The selected standard illuminant spectral power distribution

---

[1] The asterisks in L*a*b* and L*C*h° indicate the CIE 1976 system derived from XYZ. An earlier model, Hunter Lab, used similar L, a, b terminology but is based on different equations. Writing *Lab* without asterisks historically referred to Hunter Lab. The distinction is not stylistic; it reflects different mathematics.

- The CIE standard observer color-matching functions (2° or 10°)

At each wavelength, these components are multiplied together and integrated across the visible spectrum. The result is a set of tristimulus values: X, Y, and Z.

**Conceptually:**

Reflectance × Illuminant × Observer Functions (at each measured wavelength) → Integrated across all wavelengths = X, Y, Z

XYZ predicts how an average human observer would respond to that spectrum under that illuminant. Change the illuminant or observer, and the XYZ values change—even if the measured spectrum does not.

XYZ values are not perceptually uniform. They are foundational. All common print color spaces are derived from them.

---

## Why XYZ Alone Is Not Enough

Equal numeric changes in XYZ do not correspond to equal perceived color changes. This makes XYZ awkward for process control and tolerance setting.

To improve usability, XYZ is mathematically transformed into perceptually oriented spaces. The most widely used in printing is L*a*b*.

---

## L*a*b* - What the Axes Mean

L*a*b* restructures XYZ so that numeric differences more closely approximate perceived differences.

- L* represents lightness.
  *Lower values are darker. Higher values are lighter.*

- a* represents the red–green axis.
  *Positive a* moves toward red. Negative a* moves toward green.*

- b* represents the yellow–blue axis.
  *Positive b* moves toward yellow. Negative b* moves toward blue.*

L*a*b* is not a measurement. It is a transformed description derived from XYZ under defined conditions. Its validity depends entirely on those conditions.

---

## What L*a*b* Communicates Well

L*a*b* is effective for diagnosing direction of change (see Sidebar 7A):

- L* shift indicates lightness change

- a* shift indicates red–green movement

- b* shift indicates yellow–blue movement

This makes it operationally useful. When a color drifts, L*a*b* explains how it drifted.

L*a*b* is often described as *perceptually uniform*. That description is approximate. Uniformity degrades at:

- Very high chroma

- Very dark colors

- Very light colors

- Certain hue regions

L*a*b* is a model of vision, not vision itself.

# L*a*b* Color Space

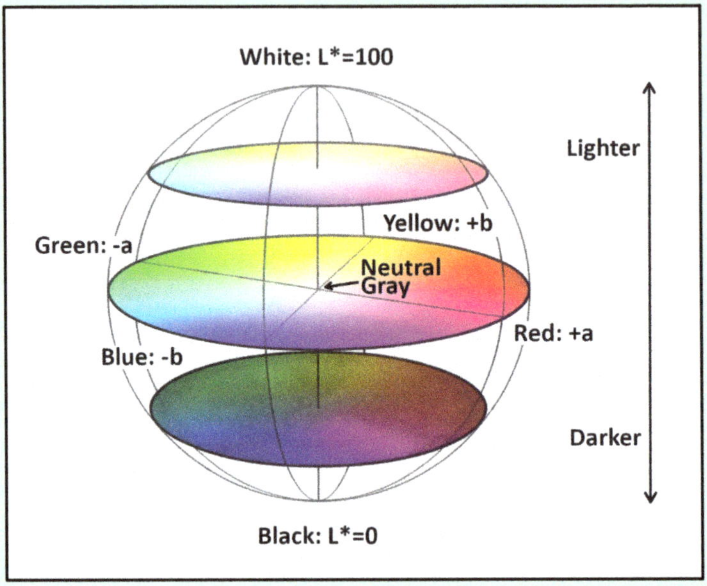


L*a*b* expresses color in a three-dimensional Cartesian coordinate system.

- L* represents lightness on a vertical axis from black (0) to white (100).

The horizontal plane is defined by two opponent axes:

- a* (– green to + red).
- b* (– blue to + yellow).

Every measured color is a point defined by three numeric values: L*, a*, and b*. Distance between two points in this space represents color difference. Because the axes are orthogonal, L*a*b* is well suited for calculating ΔE and analyzing directional shifts such as "more red," "less yellow," or "darker."

**Sidebar 7A**

# L*C*h° Color Space

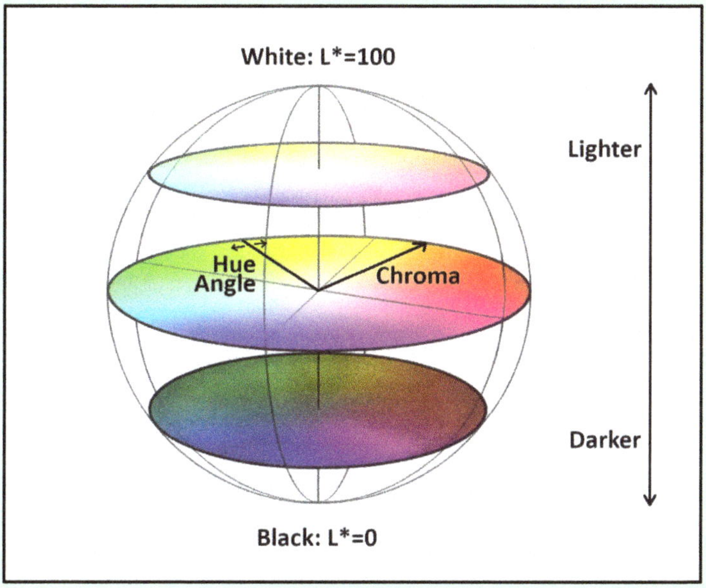


L*C*h° expresses the same color space in polar coordinates rather than rectangular form. The geometry is identical; only the coordinate description differs.

L* remains lightness.

C* (chroma) represents the radial distance from the neutral axis in the a*–b* plane. Higher values represent more chromatic (saturated) color.

h° (hue angle) represents angular position around that plane. h° indicates hue direction, measured in degrees from the +a* axis.

L*C*h° is often easier to interpret perceptually because chroma and hue are separated. However, hue angle becomes numerically unstable near the neutral axis where chroma approaches zero.

L*C*h° is not a different color space. It is a different coordinate expression of L*a*b*.

**Sidebar 7B**

## L*C*h°: A Different Representation of the Same Data

L*C*h° expresses the same color information as L*a*b*, reorganized into:

- L* (lightness, identical to L* in L*a*b*)
- C* (chroma)
- h° (hue angle)

No new information is created. It is simply a coordinate conversion (see Sidebar 7B).

- C* represents color strength. Higher values are more saturated. Lower values are more neutral.
- h° represents hue angle in the a*–b* plane. 0° aligns with the +a* axis (red). The angle increases counterclockwise, reaching 360° at the same +a* direction.

L*C*h° aligns more closely with natural language: lighter, darker, stronger, weaker, shifted in hue.

## Where L*C*h° Requires Caution

Because h° is angular, it becomes unstable near the neutral axis; when chroma is low, minor a* and b* changes can generate large numeric swings in hue angle.

For that reason, L*C*h° is excellent for visualization and explanation but must be used carefully when designing tolerances.

Remember that XYZ, L*a*b*, and L*C*h° are conditional outputs.

They depend on:

- The selected illuminant
- The selected standard observer

- The measurement geometry
- The instrument mode

If those conditions change, the numbers change.

When color numbers disagree with visual judgment, the first question is not "Is the instrument wrong?" It is *Are the conditions aligned?*

---

## Key Takeaways

- Spectral data becomes color values only after interpretation through illuminant and observer models.
- XYZ is foundational but not perceptually uniform.
- L*a*b* restructures XYZ to approximate perceptual differences.
- L*C*h° expresses the same data in intuitive terms.
- All colorimetric values are dependent on defined measurement settings, including illuminant, observer, and instrument geometry.

---

## Practical Application

If two facilities report different L*a*b* values for the same sample, verify illuminant, observer, geometry, and instrument mode before investigating process drift. In many cases, the conditions—not the samples—are different.

# Chapter 8 - Color Difference and Tolerances

## ΔE, Perception, and Production Risk

Up to this point, color has been described, measured, and expressed numerically. This chapter answers the question every production environment eventually asks: *How close is close enough?*

Color difference metrics quantify the magnitude of difference between two samples. Tolerances establish the criteria for accepting or rejecting that difference. The metric describes the difference; the tolerance determines its consequence.

This chapter explains what color difference metrics actually measure, why no single metric is perfect, and how tolerances should be structured to manage production risk rather than suggest a level of precision the process cannot consistently deliver.

---

### What a Color Difference Metric Does

A color difference metric calculates a single numeric value representing the distance between two colors within a defined color space.

That distance is intended to correlate with perceived difference. It is not a direct measure of visual judgment. It is an estimate.

The usefulness of a metric depends on how well that estimate aligns with human perception across different hues, lightness levels, and chroma values.

---

### ΔE*ab (CIE76): Simple, but Limited

ΔE*ab is an early color difference metric derived directly from L*a*b*. Each color is represented as a point defined by its L*, a*, and b* coordinates. ΔE*ab calculates the straight-line distance between two such points. The larger the distance, the greater the reported color difference.

Its strengths are simplicity and transparency. It is straightforward to calculate and easy to interpret.

Its limitation is that L*a*b* is not perfectly perceptually uniform. Equal distances in the space do not always correspond to equal perceived differences. As a result, ΔE*ab can over- or under-predict perceived difference depending on the color region.

ΔE*ab remains useful for trend analysis and basic process control, but it is a relatively coarse metric.

---

## $\Delta E_{CMC}$ and $\Delta E_{00}$ (Delta E 2000): Correcting for Perception

More advanced metrics have been developed to better align numerical differences with human visual judgment.

$\Delta E_{CMC}$ introduced perceptual weighting factors that adjust sensitivity based on lightness and chroma. It allows tolerances to be scaled differently for different types of color shifts. Developed within the textile industry, it improved correlation with visual assessments compared to ΔE*ab, particularly in applications where lightness and chroma shifts were judged unevenly.

$\Delta E_{00}$ goes further. It incorporates corrections for:

- Lightness non-uniformity
- Chroma non-uniformity
- Hue-dependent behavior
- Interactions between these components

$\Delta E_{00}$ builds on the limitations identified in both ΔE*ab and $\Delta E_{CMC}$. It applies more comprehensive perceptual corrections across the color space. While more mathematically complex, that complexity reflects the non-linear behavior of human vision.

In print and packaging applications, $\Delta E_{00}$ generally correlates better with visual judgment than ΔE*ab and often better than $\Delta E_{CMC}$ particularly for saturated colors and critical brand hues.

Other tolerance structures have been used historically, including coordinate-based limits applied independently to L*, a*, b* or to L*, C*, and h°. While these approaches were practical in earlier computational environments, they do not align well with modern perceptual difference modeling. This discussion is therefore limited to integrated color-difference metrics appropriate for contemporary print production.

---

## Why a Single Number Is Dangerous

A color difference value answers one narrow question: *How far apart are these two colors according to this model?*

It does not answer:

- Whether the difference is noticeable in context
- Whether it matters to the customer
- Whether it is consistent with previous production
- Whether it represents process drift or normal variation

Applying a single ΔE threshold as universally appropriate ignores these realities. Tolerances must be based on process capability and visual expectation, not adopted uncritically from defaults or legacy standards.

$\Delta E_{00}$ was developed to provide more uniform perceptual behavior across the color space. As a result, a single $\Delta E_{00}$ limit behaves more consistently than earlier metrics.

However, perceptual uniformity does not eliminate operational variability. Different colors carry different brand importance. Substrate and surface finish influence visual judgment. Processes differ in stability and repeatability. A tolerance must reflect these practical conditions.

A tolerance is not a physical constant. It is a decision boundary chosen to balance visual acceptability, process capability, and business risk. For that reason, tolerances should be intentionally engineered for the specific application.

---

## Tolerances Are Risk Decisions

A tolerance is not a declaration of perfection. It is a declaration of acceptable risk.

Tight tolerances reduce visual variation but increase scrap, rework, cost, and potential production disruption. Loose tolerances reduce waste but may permit unacceptable appearance shifts.

The appropriate tolerance balances:

- Visual sensitivity of the color
- Typical process variation
- Substrate and material variability
- Consequences of failure

There is no universally correct number; tolerances must reflect the realities of the process and the expectations of the customer.

---

## Why "Just Tighten the Tolerance" Fails

When color disagreements increase, the instinctive response is often to tighten tolerances.

This rarely fixes the root cause. More often, it:

- Increases false failures
- Masks configuration mismatches

- Shifts attention from process control to inspection

If a color suddenly begins failing tolerance without visible change, the correct response is to verify viewing and measurement assumptions before adjusting numbers.

If a color suddenly begins failing tolerance with visible change, determine what production variable has changed.

---

## Designing a Tolerance Strategy

Effective tolerance strategies share common characteristics

- Alignment with defined viewing conditions
- Recognition of material and substrate variability
- Consideration of normal process variation
- Periodic review informed by production data

They are documented, communicated clearly, and applied consistently.

Tolerances that exist only as software settings are not strategies. They are defaults.

---

## The Role of Visual Confirmation

Numeric tolerances do not replace visual evaluation. They help focus it.

When measurement and visual assessment agree, decisions are straightforward. When they diverge, visual judgment under controlled viewing conditions determines acceptability, and the discrepancy is investigated rather than debated.

Color control succeeds when numeric measurement and disciplined visual evaluation operate together within a clearly defined tolerance strategy.

## Key Takeaways

- Color difference metrics estimate perceived difference; they do not define acceptability.
- $\Delta E^*ab$ is simple but limited; $\Delta E_{00}$ correlates better with human perception in most print cases.
- A single numeric tolerance cannot cover all colors and situations.
- Tolerances represent risk decisions, not technical limits.
- Tightening tolerances does not fix configuration or workflow errors.
- Effective color control balances measurement with controlled visual judgment.

## Practical Application

If a long-running color suddenly begins failing tolerance without visible change, pause production adjustments. Verify illuminant assumptions, measurement modes, and reference data first. In many cases, the tolerance did not fail—the workflow did.

# Chapter 9 - How Printing Creates Color

## Inks, Substrates, Dot Structure, and Appearance Effects

Up to this point, color has been treated as light, perception, measurement, and numbers. This chapter returns color to the pressroom. Printing does not produce color by emitting light. It produces color by *modifying light through materials and structure.* That distinction explains many behaviors that appear inconsistent or unpredictable until the mechanics are understood.

Color control in printing is not about forcing materials to behave ideally. It is about understanding how they actually behave and designing workflows around those behaviors.

---

### Ink Is Not Color by Itself

Printing inks do not *add* color. They selectively absorb parts of the incident light and reflect the remainder. The color seen is the result of what survives that interaction.

Process inks are designed to be transparent so they can filter light in layers. Spot inks vary widely in transparency depending on formulation. Higher transparency allows more interaction with the substrate. Lower transparency shifts appearance toward surface color.

Because inks absorb selectively rather than absolutely, changes in ink film thickness alter the reflected spectrum. This is why density control affects appearance and why identical L*a*b* targets can be achieved through different ink film conditions with different visual consequences.

---

### Substrates Are Active Participants

Substrates are not neutral carriers. They shape color strongly.

Paper brightness, shade, texture, and fluorescence all affect the reflected spectrum. Films and foils introduce surface reflection and directional effects. Coatings modify both absorption and gloss.

Two prints using identical inks can differ visually because the substrate reflectance has changed. This is not process drift. It is physics.

This is why color targets tied to one substrate rarely transfer cleanly to another without adjustment.

---

## Dot Structure Matters More Than Many Expect

Printing creates color through patterns of dots, not continuous tones. Dot size, shape, distribution, and overlap all influence appearance.

Dot gain changes effective ink coverage, altering absorption behavior. Stochastic and conventional screening behave differently even at the same nominal tone value. Fine detail interacts with the eye's spatial resolution, changing perceived color strength.

Instruments average over structure. The eye does not always do so in the same way. This mismatch explains why two prints can measure similarly while appearing different, especially in highlights, and shadows.

---

## Overprints Create New Colors—and New Risks

When inks overlap, their combined absorption reshapes the spectrum. Overprints are not simple mixtures; they are interactive spectral events. A small shift in one ink alters the entire combined response.

Multi-ink systems increase the number of possible interactions. Even when expanded-gamut separations reduce heavy CMY overlaps in certain color regions, the system as a whole becomes more sensitive to ink density balance, gray balance, and press stability. More inks mean more interaction pathways that must remain controlled.

For this reason, expanded-gamut printing demands disciplined calibration and process control, even when targets are well defined.

Overprints are where measurement, viewing conditions, and process stability must align most closely.

---

## Gloss and Surface Effects Influence Appearance

Gloss changes how light leaves the surface. High-gloss materials increase specular reflection. Matte surfaces scatter light more diffusely.

These effects alter perceived lightness and chroma even when measured color remains stable. This is why gloss differences may cause visual disagreement without numeric confirmation.

Surface texture adds another variable. Rougher surfaces reduce effective color strength by increasing scattering. Smoother surfaces increase saturation by preserving directional reflectance.

Gloss and texture are appearance attributes, not colorimetric noise. Ignoring them leads to false conclusions.

---

## Why Printing Is Inherently Variable

Printing is a controlled process, not a static one. Ink transfer varies. Substrates vary. Environmental conditions vary. Mechanical systems drift.

Color workflows fail when they assume materials should behave like ideal mathematical objects. They succeed when they treat variation as expected and design controls to keep it within acceptable bounds.

This is why tolerances exist. It is also why tolerances must reflect material reality, not theoretical uniformity.

## Measurement Versus Appearance in Print

Instruments are essential for detecting trends and controlling process stability. They are less effective at predicting final appearance in isolation.

Measurement answers: *Is the process behaving consistently?*
Visual evaluation answers: *Does the product look acceptable under defined conditions?*

Printing requires both questions to be answered correctly.

## Designing Color Control Around Materials

Effective print color control recognizes:

- Ink behavior is spectral and thickness-dependent
- Substrates define the reference white and fluorescence response
- Surface effects may alter appearance without changing numbers

When these factors are acknowledged, disagreements decrease and efficiency improves.

## Key Takeaways

- Printing creates color by selectively absorbing and reflecting light.
- Ink transparency and film thickness strongly affect appearance.
- Substrates are active contributors to color, not passive backgrounds.
- Dot structure and screening influence perception beyond averaged measurements.

- Overprints increase interaction sensitivity and require disciplined control.
- Gloss and texture change appearance even when measurements are stable.
- Printing variation is normal and must be managed, not denied.

---

## Practical Application

If a color meets numeric targets but appears weaker on a new substrate, do not immediately adjust tolerances or blame the press. Evaluate substrate reflectance, fluorescence, and surface effects first. In many cases, the color system is behaving correctly, and the target needs to be redefined for the material.

# Chapter 10 - ICC and Proof-to-Press Alignment

## Intent, Assumptions, and Common Failure Modes

By now, the physical and perceptual realities of print color are established. Color depends on light, materials, human vision, and controlled measurement. ICC color management connects those realities to digital workflows through a structured translation system. When device profiles, measurement conditions, and viewing assumptions reflect physical reality, proof and press remain aligned. When those assumptions drift from the process, alignment degrades—not because the math is wrong, but because the inputs no longer represent the materials and conditions on press.

This chapter explains what ICC systems are designed to do, what assumptions they rely on, and why proof-to-press alignment breaks when those assumptions are ignored.

---

### What ICC Color Management Is—and Is Not

ICC color management is a translation system. It maps color values from one device-dependent space to another using profiles that describe how each device behaves under defined conditions.

- It is not a color correction system.
- It is not a guarantee of visual match under all conditions.
- It is not a replacement for viewing discipline or process control.

ICC workflows work best when treated as controlled transformations governed by defined assumptions, not as self-correcting systems.

## Profiles Describe Behavior Under Assumptions

An ICC profile describes how a device reproduces color under a defined set of conditions. Those conditions include:

- A defined illuminant and observer
- A defined reference white and substrate
- A defined measurement configuration and geometry
- A defined ink set and ink film behavior
- A defined screening method and tone reproduction
- A defined, stabilized state of the press or output device

In print workflows, that stabilized state may reflect specific substrate characteristics, ink formulations, press settings, screening strategy, and mechanical configuration at the time of characterization.

Change those conditions, and the profile no longer describes the device's behavior accurately.

This is the most common source of misalignment. Profiles are often reused long after the materials, press behavior, or viewing assumptions that created them have changed.

## Source, Destination, and Intent

Every ICC conversion has at least three components:

- A source profile
- A destination profile
- A rendering intent

The source profile defines what the numbers mean. The destination profile defines what the device can produce. The rendering intent defines how compromises are handled when colors cannot be reproduced exactly.

In contract proofing workflows, the proof is not merely rendered through the proofer's native profile. It incorporates a conversion through the press profile so the proof simulates the press's characterized behavior under defined conditions. The press profile functions as the predictive model; the proofing device is used to visualize that model as accurately as possible.

Misunderstanding any one of these leads to incorrect expectations.

---

## Rendering Intents Are Trade-Offs

Rendering intents do not "improve" color. They choose which errors are acceptable.

- **Perceptual** redistributes color relationships to preserve overall appearance.
- **Relative colorimetric** preserves in-gamut colors and maps out-of-gamut colors to the closest reproducible color, while shifting the source white point to match the destination paper white .
- **Absolute colorimetric** preserves the source white point, including its color cast, relative to the destination white point, which can allow paper-white simulation for proofing.
- **Saturation** prioritizes vividness over accuracy.

No intent is universally correct. The correct choice depends on the purpose of the reproduction and the approval context.

---

## Proofs Are Simulations, Not Targets

A contract proof is a simulation of expected press appearance under defined conditions. It predicts how the final print should look when viewed under specified lighting and evaluation standards. It is not a physical replication of press behavior.

A proof should not be treated as a numeric target to be measured and matched exactly. The proofing device, substrate, ink set, and screening method differ from those of the press. Even when profiles are accurate, the two systems achieve similar appearance through different physical mechanisms.

Proof-to-press alignment fails when proofs are treated as absolute numeric references instead of conditional predictions. Comparing measurements between proof and press without accounting for those differences can produce apparent failures even when the visual match is acceptable.

The purpose of a proof is visual agreement under controlled viewing conditions—not strict numeric equivalence.

---

## Viewing Conditions Are Part of ICC Workflows

ICC profiles assume specific viewing conditions. If proofs are evaluated under uncontrolled lighting or presses are judged under different illumination, the system breaks even if the profiles are correct.

This is why viewing discipline precedes profiling. Profiles cannot compensate for inconsistent evaluation.

---

## Common Proof-to-Press Failure Modes

Several failure patterns repeat across organizations:

- Proofs approved under one illuminant, press judged under another
- Profiles built with one measurement configuration, verification done with another

- Substrate substitutions made without updating expectations
- Profiles reused after process changes
- Numeric proof-to-press comparisons used instead of visual confirmation

In each of these cases, the issue lies not with ICC mathematics, but with inconsistent workflow assumptions.

---

## ICC Is a Framework, Not a Workflow

ICC color management provides the framework for translating color. It does not define how approvals are made, how tolerances are set, or how disputes are resolved.

Those decisions belong to the workflow. When ICC is treated as plug-and-play, it becomes a source of confusion rather than control.

Successful programs define:

- When ICC is authoritative
- When visual judgment prevails
- How disagreements are resolved

---

## Aligning Digital Intent With Physical Reality

The most reliable proof-to-press workflows align digital assumptions with physical behavior:

- Profiles reflect current press conditions
- Proofing systems are validated visually, not just numerically
- Substrate differences are acknowledged explicitly
- Approvals are tied to defined viewing conditions

When this alignment exists, ICC systems perform predictably. When it does not, no amount of profile tuning fixes the underlying mismatch.

ICC profiling succeeds when it is treated as a disciplined translation framework embedded within a disciplined workflow.

---

## Key Takeaways

- ICC color management translates color between devices under defined assumptions.
- Profiles describe behavior; they do not enforce correctness.
- Rendering intents represent trade-offs, not quality levels.
- Proofs are visual simulations, not numeric targets.
- Viewing conditions are integral to ICC workflows.
- Most ICC failures are workflow failures, not technical ones.

---

## Practical Application

If a press sheet fails to *match the proof* numerically but looks correct under the approved viewing conditions, do not adjust the press to chase numbers. Reconfirm the role of the proof as a visual reference and verify that evaluation conditions align with the ICC assumptions. In many cases, the system is functioning as designed.

# Chapter 11 - A Repeatable Color Decision Workflow

## From Design Intent to Press Control

This book has deliberately avoided shortcuts. Color has been built from its physical and perceptual foundations: light, spectra, human vision, viewing conditions, measurement, color spaces, difference metrics, materials, and digital translation. The final step is integration.

Predictable color does not come from any single tool or number. It comes from a workflow that defines *who decides what, under which conditions, using which references*. This chapter assembles those elements into a repeatable color decision workflow suitable for real production environments.

---

### The Core Principle

Every color decision must answer three questions:

- What is the visual reference?
- Under what conditions is it valid?
- How is conformance verified?

If any are undefined, disagreement is inevitable.

---

### Step 1: Define Design Intent Clearly

Color intent must be explicit. Vague language such as *match the proof* or *match last time* is not intent.

Effective intent definition includes:

- The reference (proof, standard, previous production, or approved sample)
- The substrate and print process

- The viewing conditions under which approval applies

Intent definition is a human task. No instrument can infer it.

---

## Step 2: Lock Viewing Conditions Before Evaluation

Before any approval or comparison:

- The viewing illuminant is defined
- Surrounds are neutral
- Observers are adapted
- Ambient light is controlled

This step is non-negotiable. Skipping it invalidates every decision that follows.

---

## Step 3: Establish Measurement Assumptions

Measurement assumptions must be defined and documented:

- Illuminant
- Observer
- Geometry
- Measurement mode
- Specular handling
- Aperture
- Backing material

These settings define what the numbers represent. Changing them mid-stream resets the workflow.

## Step 4: Use Measurement to Stabilize, Not Override

Measurement is used to:

- Detect process drift
- Compare samples objectively
- Support visual decisions

It is not used to replace visual approval. When numbers and vision disagree under controlled conditions, vision determines acceptability, and the discrepancy is investigated to identify its source.

## Step 5: Apply Tolerances as Risk Controls

Tolerances are applied based on:

- Visual sensitivity of the color
- Material and substrate variability
- Demonstrated process capability
- Consequence of failure

Many organizations establish a standard shop tolerance for efficiency and consistency. This can provide a useful baseline. However, not all colors and customers carry equal risk. Critical brand colors may justify tighter limits, while less sensitive work may allow broader ones.

A single global threshold may simplify administration, but it rarely reflects the full range of visual and commercial expectations across a workflow.

Tolerances should be reviewed periodically against production data and customer requirements. They are not fixed constants; they are decision limits that evolve with process performance and business context.

## Step 6: Manage Material Changes Explicitly

Any change in:

- Substrate
- Ink formulation
- Coating or finish
- Lighting environment

is treated as a workflow event, not a surprise. Expectations are reset, references are updated if needed, and approvals are reconfirmed.

Material reality must be acknowledged, not ignored.

---

## Step 7: Align Digital and Physical Systems

Digital color management is aligned to physical behavior:

- Profiles reflect current conditions
- Proofs are validated visually
- Rendering intents are chosen intentionally
- Numeric comparisons respect assumptions

Digital systems support decisions; they do not dictate them.

---

## Step 8: Resolve Disagreements Systematically

When disagreement occurs, the resolution sequence is fixed:

1. Verify viewing conditions
2. Verify measurement assumptions
3. Verify reference validity

4. Compare visual appearance under defined conditions
5. Adjust targets or expectations only if necessary

Disagreement is replaced with diagnostics.

---

## Step 9: Document and Communicate

A workflow only works if it is shared.

Documentation includes:

- Approved viewing conditions
- Measurement configurations
- Tolerance logic
- Approval responsibilities

Communication ensures that production, prepress, quality, and approval teams operate from the same assumptions.

---

## What This Workflow Prevents

When followed, this workflow prevents:

- Chasing numbers without visual justification
- Re-approving color under uncontrolled lighting
- Blaming instruments for configuration errors
- Treating material behavior as failure
- Escalating subjective disputes

It does not eliminate variation. It makes variation manageable.

## What This Workflow Enables

This workflow enables:

- Faster approvals
- Fewer disputes
- More stable production
- Clear accountability
- Defensible decisions

It allows teams to spend time improving results instead of debating causes.

## Final Perspective

Color in printing will never be perfectly predictable. Materials vary. People adapt. Conditions change.

Predictable color does not mean eliminating variability. It means *controlling the variables that matter, acknowledging the ones that cannot be eliminated, and making decisions under shared assumptions.*

A disciplined workflow turns color from a source of disagreement into a managed engineering outcome.

## Key Takeaways

- Predictable color comes from workflows, not tools.
- Viewing conditions define the visual truth.
- Measurement stabilizes decisions when assumptions are aligned.
- Tolerances manage risk, not perfection.

- Material behavior must be designed into expectations.
- Disagreements should trigger diagnostics, not debate.
- A shared workflow is the most powerful color control tool available.

---

## Practical Application

If your organization experiences frequent color disagreements, do not start by changing targets or buying new tools. Map your current decisions against this workflow and identify where assumptions are undefined or inconsistent. Correcting those gaps often delivers immediate improvement without changing a single color value.

# Glossary

Below are workflow-focused definitions for the core terms used in this book. I have kept the definitions short, assuming industry literacy. They are designed to align interpretation, not teach fundamentals.

---

**45°/0° Geometry**

A measurement geometry in which the sample is illuminated at 45 degrees and measured at 0 degrees. It correlates well with visual appearance and is widely used for press-side control.

---

**a***

The red–green axis in L*a*b*. Positive values indicate movement toward red; negative values indicate movement toward green.

---

**Absorption**

The selective removal of specific wavelengths of light by a material. Printing inks create color primarily through absorption.

---

**Adaptation (Visual Adaptation)**

The continuous adjustment of the visual system to prevailing lighting and recent visual history. Stabilizing adaptation is necessary before reliable color evaluation.

---

**Aperture**

The diameter of the measurement area of a spectrophotometer. Aperture size affects how much surface variability is averaged into a reading and must be appropriate for the substrate texture and patch size.

**b***

The yellow–blue axis in L*a*b*. Positive values indicate movement toward yellow; negative values indicate movement toward blue.

---

**Calibration**

The process of aligning an instrument to a reference standard to ensure accurate measurement.

---

**Chroma (C*)**

The perceived strength or saturation of a color relative to neutral in the L*C*h° color space.

---

**Chromatic Adaptation**

The visual system's ability to partially discount the color of the illumination so objects appear stable under different lighting conditions.

---

**Color Memory**

The limited ability of human observers to recall the exact appearance of a color once it is removed from view. Because color memory is unreliable even over short intervals, accurate color evaluation requires samples to be viewed side-by-side under controlled conditions.

---

**Correlated Color Temperature (CCT)**

A numerical description of the apparent warmth or coolness of a light source. It does not fully describe spectral structure.

**d/8° Geometry**

A measurement geometry that uses diffuse illumination inside an integrating sphere and measures at an 8-degree angle. It is preferred for textured, metallic, or specialty ink and substrates. Measurements may be taken with specular reflection (component) included (SCI), or Excluded (SCE).

---

**ΔE*ab (CIE76)**

The original Euclidean color difference formula calculated in L*a*b* color space. While simple to compute, it does not fully correlate with human perception and is increasingly replaced by $\Delta E_{00}$ in printing workflows.

---

**$\Delta E_{CMC}$**

A color difference metric applying perceptual weighting factors to improve correlation with visual assessment.

---

**$\Delta E_{00}$ (ΔE2000)**

A color difference formula designed to better align numerical differences with human visual perception. It accounts for interactions between lightness, chroma, and hue and is the dominant metric used in modern printing color standards.

---

**Expanded Color Gamut (ECG)**

A printing approach that uses additional process inks to extend the reproducible color space beyond CMYK. ECG reduces reliance on spot colors but requires disciplined color management. This is sometimes referred to as *Extended Color Gamut.*

---

**Fluorescence**

Emission of visible light after absorption of shorter-wavelength energy, typically ultraviolet. OBAs exhibit fluorescence.

**Hue Angle (h°)**

The angular position of a color in L*C*h° space that defines its color family. Small hue angle shifts can be visually significant even when overall color difference values appear small.

**ICC Profile**

A data file that characterizes how a device reproduces color. ICC profiles enable color-managed workflows by translating color data consistently between devices and processes.

**Illuminant**

The simulated light source used in color measurement calculations. Changing the illuminant alters reported values and can reveal metamerism between samples.

**Ink Film Thickness**

The thickness of ink deposited on the substrate during printing. Ink film thickness strongly influences chroma, lightness, and sometimes hue, making it a primary press control variable.

**L***

The lightness coordinate in L*a*b*, representing relative brightness from black to white.

**L*a*b* (CIELAB)**

A device-independent color space defined by the CIE that represents color using lightness (L*), red–green (a*), and yellow–blue (b*) axes. It is the primary language for specifying and evaluating color in printing.

### L*C*h° (CIELCH)

A polar representation of L*a*b* expressing lightness (L*), chroma (C*), and hue angle (h°). L*C*h° is sometimes preferred for diagnosing color errors because it separates strength and hue effects more clearly.

### M0

A legacy measurement mode based on unfiltered illumination. M0 remains common due to historical data but may produce inconsistent results on OBA-containing substrates when comparing data from instruments manufactured by different vendors.

### M1

A measurement mode designed to accurately simulate D50 illumination with controlled UV content. M1 is the recommended standard for modern printing workflows. Note: When neither the substrate nor the ink contains optical brighteners, measurements from M0, M1, and M2 should match.

### M2

A measurement mode that uses a UV-cut illumination to eliminate the impact of optical brightening agents. M2 is commonly used for building ICC profiles.

### M3

A measurement mode that uses polarized illumination to reduce the influence of surface gloss. M3 may be useful on glossy prints, films, and varnished surfaces where specular reflection can affect readings, but is almost never used for press-side color control.

**Measurement Geometry**

The configuration by which a spectrophotometer illuminates and observes a sample. Geometry significantly affects measurement results and must be consistent across workflows.

**Measurement Mode**

Standards for 45°/0° spectrophotometer illumination and filtration that are particularly relevant when substrates contain optical brightening agents. The measurement mode must match how the color standards were created.

**Metamerism**

A phenomenon in which two samples match under one light source but differ under another. Metamerism results from differences in spectral reflectance.

**Optical Brightening Agents (OBAs)**

Fluorescent compounds added to substrates to increase perceived whiteness. OBAs influence both appearance and measurement and require controlled measurement conditions. OBAs are present in almost all proofing stocks and are occasionally added to inks.

**Overprint**

The interaction of overlapping inks, producing a new spectral response through combined absorption.

**Perceptual Uniformity**

The degree to which equal numeric differences in a color space correspond to equal perceived differences.

**Rendering Intent**

A strategy within ICC workflows defining how out-of-gamut colors are handled during color conversion.

**Repeatability**

The ability of an instrument and procedure to produce consistent results under unchanged conditions.

**SCI (Specular Component Included)**

A d/8° measurement condition in which specular reflection is included in the measurement. Specular included data represents the intrinsic color of a material and minimizes the influence of surface gloss.

**SCE (Specular Component Excluded)**

A d/8° measurement condition in which specular reflection is removed from the measurement. Specular excluded data correlates more closely with perceived appearance and is influenced by surface gloss and texture.

**Simultaneous Contrast**

A perceptual phenomenon in which adjacent colors influence each other's perceived hue, lightness, or chroma.

**Spectral Power Distribution (SPD)**

A description of the relative energy output of a light source at each wavelength across the visible spectrum.

**Spectral Reflection**

Light that is reflected from a surface at an angle equal to the angle of incidence, producing mirror-like reflection. Specular reflection is strongly influenced by surface gloss and texture.

---

**Spectrophotometer**

An instrument that measures spectral reflectance, transmittance, or emission and derives colorimetric values under defined conditions.

---

**Standard Observer**

A mathematically defined model of average human color vision used in colorimetric calculations.

---

**Substrate**

The material on which printing occurs, such as paper, film, or foil.

---

**Tolerance**

The acceptable range of deviation from a color target. Effective tolerances reflect both visual sensitivity and process capability.

---

**XYZ (CIEXYZ)**

Tristimulus values calculated from spectral data using a defined illuminant and standard observer. XYZ is the foundation of modern colorimetry.

## About the Author

Bob Binder is a retired printing and color-management professional who now works as an independent consultant through RGBinder, LLC.

Bob began his career in 1978 in Grand Rapids, Michigan, working as a film stripper at Acraforms while earning a degree in Printing Management from Western Michigan University. After graduation, he moved to Evansville, Indiana, where he served as Prepress Manager at Koch Label Company in a gravure label production environment. He later relocated to Carol Stream, Illinois, and spent ten years as Prepress Manager at Jefferson-Smurfit Corporation, a folding-carton facility operating gravure and offset presses.

These roles provided hands-on experience in prepress, pressroom production, and cross-process color control.

Bob later returned to Grand Rapids and spent 26 years in customer-facing technical and support roles, most recently as a Solution Architect at X-Rite. His work expanded into color theory, color measurement, instrument geometry, standards development, workflow integration, and deployment of color-control systems across printing environments. He has had technical articles published in *The Big Picture* and *Flexo* Magazine.

Throughout his career, Bob has worked closely with press operators, quality managers, and production teams. That exposure shaped his practical approach to color science: measurement must connect to vision; theory must connect to application; and standards must be understood before they are trusted.

This book reflects that philosophy, building color understanding from fundamentals—light, spectra, human vision, measurement assumptions, and translation models—so readers can interpret numbers accurately and apply them with confidence.

Additional resources and video demonstrations are available at: *rgbinder.com/colortheorybook*

www.ingramcontent.com/pod-product-compliance
Ingram Content Group UK Ltd.
Pitfield, Milton Keynes, MK11 3LW, UK
UKHW022028041026
15591UKWH00012B/562

9 798994 772911